An-Nan Chen

Selective Laser Sintering

An-Nan Chen

Selective Laser Sintering

A strategy for manufacturing complicated-structured ceramic foams using fly ash hollow spheres

LAP LAMBERT Academic Publishing

Publisher:
LAP LAMBERT Academic Publishing
is a trademark of
International Book Market Service Ltd., member of OmniScriptum Publishing Group
17 Meldrum Street, Beau Bassin 71504, Mauritius

Printed at: see last page
ISBN: 978-620-0-53035-6

Selective Laser Sintering (SLS): A strategy for manufacturing complicated-structured ceramic foams using fly ash hollow spheres

An-Nan Chen*, Jia-Min Wu, Yu-Sheng Shi

State Key Laboratory of Materials Processing and Die & Mould Technology, School of Materials Science and Engineering, Huazhong University of Science and Technology, Wuhan 430074, China

* Corresponding author. Tel: +86 27 87558155; fax: +86 27 87558155.
E-mail address: AnnanChenNUAA@hust.edu.cn (A.-N. Chen)

Acknowledgements

I am grateful to my supervisors Prof. Yu-Sheng Shi and Dr. Jia-Min Wu for their guidance and help through the years of scientific researches. They have been strong and supportive advisors to me and have always given me great freedom to pursue independent research work. I would also like to express my gratitude to Dr. Li-Jin Cheng and Dr. Rong-Zhen Liu for their suggestions and good knowledge in characterization of ceramic materials.

Sincere thanks to the members of our research group (Rapid Manufacturing Center in Huazhong University of Science and Technology (HUST) of China), Prof. Chen-hui Li, Dr. Shi-Feng Wen for the provision of manufacturing equipment and technical support. I would also like to thank to Prof. Qing-Song Wei, Prof. Chun-Ze Yan, Prof. Bo Song and Dr. Jie Liu for suggestions.

I would like to thank my co-authors Miss Meng Li and Mr Lu Lu for fruitful cooperation among theoretical and practical matters. I would like to express my gratitude to Dr. Ying Chen, Dr. Yan Zhou and Dr. Jin Su for discussions and suggestions through this work.

Very special thanks to Wuhan Huake 3D Technology Co., Ltd. for co-operation and partnership.

Major financial support has been received from the National Natural Science Foundation of China (51605177), National Key R&D Program of

China (2018YFB1105503) and Fundamental Research Funds for the Central Universities (2018KFYYXJJ030).

For all my friends in HUST thank you for all the countenance.

I would like to thank my parents and for all their love, patience and understanding, they always prayed for the success of my study. I most want to thank my wife Ms Zheng-Kang Xu. It was my wife's care, support and encouragement that inspired me to move forward optimistically and positively.

I am very sorry that I am not able to mention all people who have given a contribution to this study. I am sure that some people will be missing to be mentioned. I would like to thank all of you.

Table of content

1 Ceramic Additive Manufacturing

Ceramic materials have attracted considerable attentions over the past decades due to their impressive performance, including high mechanical strength, good thermal and chemical stability, and outstanding electrical and magnetic properties etc. [1]. Typically, the ceramic object are manufactured into a desired shape staring from a mixture of ceramic powders, binders and stabilizers, using conventional forming techniques such as die pressing, injection moulding, slip casting, tape casting and gel casting [2]. The further densification is achieved through high temperature sintering of the green object. Nonetheless, these conventional forming techniques have some restrictions set by geometrical design in fabricating objects with high structural complexity, for example, the honeycomb ceramics with interconnected holes which are now widely used in exhaust filtration [3]. On the other hand, the machining of ceramic objects is extremely difficult due to their extreme hardness and brittleness, which would easily wear cutting tools and cause cracks in ceramic objects.

Additive Manufacturing (AM), also known as 3D printing developed in the late 1980's, describes a class of techniques in which a three-dimensional object is directly fabricated in a layer-by-layer additive manner from a computer aided design (CAD) data without using a preform or mold [4]. In view of the benefits of short production cycles and the ability that enables the flexible preparation of high geometrical complexity and precise structures, the AM technique have become nowadays part of the advanced forming process that are widely applied to the near-net-shape fabrication of metallic, polymeric and ceramic materials [5-7]. In comparison with metals and polymers, the brittle ceramic materials have more challenges to AM process by virtue of their extremely high melting temperature, low thermal shock resistance and low or no plasticity [8]. To date, with the latest advances in materials science and computer science, AM techniques that have been widely used for ceramic manufacturing including laminated object manufacturing [9], fused deposition modelling [10], inkjet printing [11], stereolithography [12] and selective laser sintering/melting (SLS/SLM) [13]. As a consequence, ceramic additive manufacturing has nowadays rapidly gained increasing attention across the science and engineering communities.

1.1 Selective laser sintering (SLS)

Selective laser sintering (SLS), known as one of the powder-based AM technologies developed in 1980s, is considered as a promising way to fabricate complex-shaped ceramic objects by selectively sintering powders layer by layer using a high-intensity laser beam (e.g. CO_2 laser) [14]. The schematic of SLS process can be seen in Figure 1.1. Typically, the SLS building process mainly consists of the powder deposition and laser sintering. As for the powder deposition, a conventional roller or scraper system is utilized to deposit powder layers which are then locally heated and selectively sintered according to the predefined geometries using the laser beam. A heating system is used to preheat the powder bed during SLS which can reduce thermal stresses and thus help prevent crack formation in sintered objects [15].

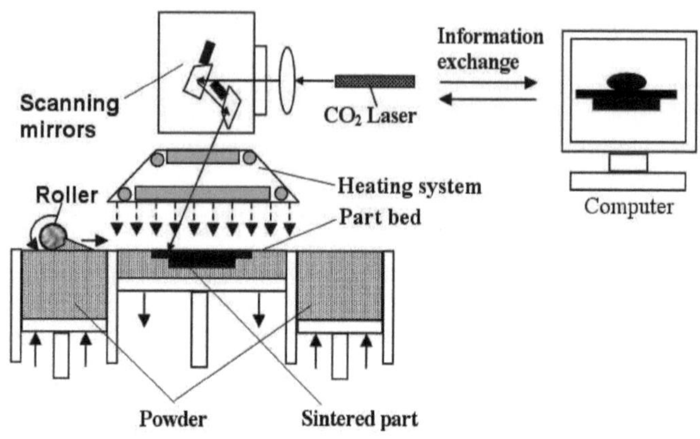

Figure 1.1 Schematic of the SLS process [13]

Generally, SLS of ceramic objects could be achieved mainly by two approaches, i.e., directly and indirectly. In the case of direct SLS, ceramic powders are directly heated and melted by the laser irradiation, while the laser sintering process is generally complicated due to the poor thermal shock resistance of ceramic materials. Furthermore, the sintered objects always show rather low density and poor mechanical properties, which is attributed to the insufficient material diffusion resulting from the high heating and cooling rates during laser irradiation [16]. To improve the density and performance of the direct sintered ceramic objects, a multitude of techniques have been developed and introduced such as pretreated pressing [17] and self-propagating high-temperature synthesis (SHS) [18]. These studies, to some extent, have proposed several feasible approaches to manufacture relative dense ceramic objects by direct SLS. However, easy-cracking is still difficult to avoid in the fabricated objects due to the huge thermal stresses, which would result in the poor mechanical properties. The indirect SLS (iSLS), by introducing the low-melting sacrificial binder phase, could alternatively be an appropriate way to fabricate crack-free objects by sintering binders that can easily fuse ceramic powders together to obtain the green objects. Nonetheless, it is not easy to densify the iSLSed ceramic objects during furnace sintering due to the low packing density of green objects, and thus the mechanical strength of ceramics is generally rather poor for actual applications. In order to overcome the low density and poor mechanical properties of ceramic objects prepared by indirect SLS, lots of research have been done by Prof. Shi Yu-Sheng's group in Huazhong University of Science and Technology (HUST). A hybrid technology named SLS/Cold Isostatic Pressing (CIP) was firstly proposed by Shi's group in 2010 [19], which introduced the CIP technology into the field of SLS to fabricate ceramic objects with improved theory density. The schematic of SLS/CIP process is shown in Figure 1.2.

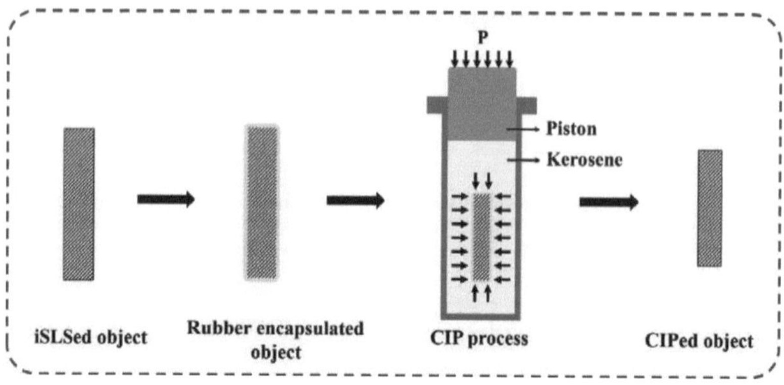

Figure 1.2 The schematic of SLS/CIP process [20]

Taking the Al_2O_3/Epoxy resin (E06) samples as example [21]: Figure 1.3 shows the microstructure of fracture surface of green Al_2O_3/E06 objects prepared by iSLS followed by CIP process. It can be seen from Figure 1.3(a) that the alumina agglomerates are bonded together through the bonding necks formed by the melted E06 after iSLS (shown in Figure 1.3(b)). Quantities of pores (relative density of 33%) could be easily observed in green objects suggesting a poor densification process during laser irradiation. After CIP post-processing, as shown in Figure 1.3(c) and (d), no obvious pores can be observed in green objects and the alumna particles are squeezed together, which results in a high relative density of more than 92% after furnace sintering. The results indicate that SLS/CIP technology could be the promising route for preparing complex ceramic objects with enhanced mechanical performance.

Figure 1.3. (a-b) Microstructure of fracture surface of Al_2O_3/E06 samples prepared by iSLS and (c-d) followed by CIP process with dwell pressure of 335 MPa [21]

In recent years, various ceramic objects such as Al_2O_3, ZrO_2, SiO_2, SiC [13], have already been prepared using the iSLS of ceramic-binder composites as shaping step. Previous studies have mainly focused on the density improvement to obtain compact ceramic objects, however, reports have rare on SLS technique to prepare porous ceramics. In fact, it is notably that the iSLS technique has unique advantages in building porous objects as the powder layers are generally not compacted during layer deposition, and the sacrificial binder phase would be burnt out under elevated temperature that generate open pores in ceramic objects.

1.2 Ceramics materials and binders for SLS

In general, the particle size of ceramic powders used for laser sintering are required to be in a micron level with spherical shapes, which possess good flowability that can be easily deposited using a roller or scraper. Therefore, the following two methods in the selection of ceramic powders for SLS are commonly taken into consideration: one is using the micron-sized powders obtained by granulation of nanoscale or submicron powders, the other is covering binders on the surface of ceramic powders. In addition, the inherent physical and chemical properties of materials such as thermal absorptivity and conductivity, melt viscosity and surface tension, crystallization temperature and velocity, all have a significant influence on the quality of SLS fabricated objects, which should be fully considered when selecting ceramic powders for SLS [22].

To improve the microstructure and composition homogeneity of SLSed objects, the binders typically should meet the requirements of low melting point, low liquid viscosity and good wettability with the substrate ceramic materials, which would promote the diffusion and migration of materials after melting at high temperature. At present, there are mainly three kinds of binders used in the formation of ceramic materials through iSLS, i.e., the inorganic binders such as $NH_4H_2PO_4$ and B_2O_3, the organic binders such as epoxy resin and phenolic resin, and the metal binders such as Al powders, etc. [23]. Furthermore, the binder content added into the compositions also has a great influence on the quality of iSLSed objects. When the binder addition is insufficient, some defects, such as cracking and delamination, would be observed easily in green bodies due to the incomplete adhesion between ceramic matrix particles. However, excessive binders result in the relative low volume fraction of ceramic powders, which would lead to the large shrinkage during the subsequent binder removal and furnace sintering processes, easily causing the deformation and cracking. Therefore, the amount of binders added into the compositions for SLS, without affecting the final density and strength of prepared objects, is as less as possible.

1.3 Advantages of SLS-formed ceramic foams using FAHSs

It is mentioned that the iSLS has inherent advantages in building porous objects considering the incompact powder deposition and the sacrificial binder phase. However, the mechanical performance of the prepared porous objects with high porosity probably require further improvement, and the issues of large shrinkage and deformation needs to be solved. In this study, we proposed a feasible strategy for manufacturing complicated-structured ceramic foams by iSLS using fly ash hollow spheres (FAHSs) as raw materials. The FAHSs are inartificial with hollow nature and also the by-product with ultra-low cost generated from combustion process in coal power plants, which have caused serious environmental pollution [24]. The major superiority of using FAHSs as raw materials to prepare ceramic foams can be listed as follow:

(i) their particle size and distribution can be well adjusted before use, which can control the pore size and porosity of ceramic objects;

(ii) their shell walls can act as skeletons, which might potentially improve mechanical properties of mullite foams;

(iii) they are ultra-low cost, and the recycling of such solid waste can help solve pollution problem caused by FAHSs.

References

[1] Dutta S. Fracture toughness and reliability in high-temperature structural ceramics and composites: prospects and challenges for the 21st century. Bulletin of Materials Science, 2001, 24(2): 117-120.

[2] Yu F., Wang H., Bai Y., et al. Preparation and characterization of porous Si_3N_4 ceramics prepared by compression molding and slip casting methods. Bulletin of Materials Science, 2010, 33(5): 619-624.

[3] Du L., Liu W., Hu S., et al. Preparation and photocatalytic properties of macroporous honeycomb alumina ceramics used for water purification. Journal of the European Ceramic Society, 2014, 34(3): 731-738.

[4] Sing S.L., Yeong W.Y., Wiria F.E., et al. Characterization of titanium lattice structures fabricated by selective laser melting using an adapted compressive test method. Experimental Mechanics, 2016, 56(5): 735-748.

[5] Gu D.D., Meiners W., Wissenbach K., et al. Laser additive

manufacturing of metallic components: materials, processes and mechanisms. International materials reviews, 2012, 57(3): 133-164.

[6] Salmoria G.V., Leite J.L., Vieira L.F., et al. Mechanical properties of PA6/PA12 blend specimens prepared by selective laser sintering. Polymer Testing, 2012, 31(3): 411-416.

[7] Travitzky N., Bonet A., Dermeik B., et al. Additive manufacturing of ceramic‐based materials. Advanced Engineering Materials, 2014, 16(6): 729-754.

[8] Shahzad K., Deckers J., Boury S., et al. Preparation and indirect selective laser sintering of alumina/PA microspheres. Ceramics International, 2012, 38(2): 1241-1247.

[9] Zhong H., Yao X., Zhu Y., et al. Preparation of SiC ceramics by laminated object manufacturing and pressureless sintering. J Ceram Sci Technol, 2015, 6: 133-140.

[10] Corcione C.E., Gervaso F., Scalera F., et al. Highly loaded hydroxyapatite microsphere/PLA porous scaffolds obtained by fused deposition modelling. Ceramics International, 2019, 45(2): 2803-2810.

[11] Derby B. Additive manufacture of ceramics components by inkjet printing. Engineering, 2015, 1(1): 113-123.

[12] Halloran J.W. Ceramic stereolithography: additive manufacturing for ceramics by photopolymerization. Annual Review of Materials Research, 2016, 46: 19-40.

[13]   Chen A.N., Wu J.M., Liu K., et al. High-performance ceramic parts with complex shape prepared by selective laser sintering: a review. Advances in Applied Ceramics, 2018, 117(2): 100-117.

[14]   Beaman J.J., Deckard C.R. Selective laser sintering with assisted powder handling: U.S. Patent 4,938,816. 1990-7-3.

[15]   Wilkes J.I. Selektives laserschmelzen zur generativen herstellung von bauteilen aus hochfester oxidkeramik. PhD thesis, RWTH Aachen University, Aachen, Germany, 2009.

[16]   Zocca A., Colombo P., Gomes C.M., et al. Additive manufacturing of ceramics: issues, potentialities, and opportunities. Journal of the American Ceramic Society, 2015, 98(7): 1983-2001.

[17]   Liu R.Z., Wu J.M., Chen A.N., et al., Balling phenomenon and cracks in alumina ceramics prepared by direct selective laser melting assisted with pressure treatment. Ceramics International, 2020.

[18]   Gahler A., Heinrich J.G., Guenster J. Direct laser sintering of Al_2O_3-SiO_2 dental ceramic components by layer-wise slurry deposition. Journal of the American Ceramic Society, 2006, 89(10): 3076-3080.

[19]   Liu J., Zhang B., Yan C., et al. The effect of processing parameters on characteristics of selective laser sintering dental glass-ceramic powder. Rapid Prototyping Journal, 2010, 16(2): 138-145.

[20]   Chen F., Wu J.M., Wu H.Q., et al. Microstructure and mechanical properties of 3Y-TZP dental ceramics fabricated by selective laser

sintering combined with cold isostatic pressing. International Journal of Lightweight Materials and Manufacture, 2018, 1(4): 239-245.

[21]　Liu K., Shi Y., Li C., et al. Indirect selective laser sintering of epoxy resin-Al_2O_3 ceramic powders combined with cold isostatic pressing. Ceramics International, 2014, 40(5): 7099-7106.

[22]　Kolosov S., Vansteenkiste G., Boudeau N., et al. Homogeneity aspects in selective laser sintering (SLS). Journal of materials processing technology, 2006, 177(1-3): 348-351.

[23]　Kruth J.P., Wang X., Laoui T., et al. Lasers and materials in selective laser sintering. Assembly Automation, 2003, 23(4): 357-371.

[24]　Hu J., Chen M., Fang X., et al. Fabrication and application of inorganic hollow spheres. Chemical Society Reviews, 2011, 40(11): 5472-5491.

2 FAHSs/PA12 composites and ceramic foams preparation

In general, there are mainly four kinds of ceramic materials used for iSLS that have been extensively investigated, including ceramic powders directly mixed or coating with binders, surface modification, and the resin sand [1]. Furthermore, the ceramic/binder compositions suitable for iSLS can be prepared by technologies of mechanical mixing, spray granulation and surface coating [2], which involves the approaches such as mechanical mixing, solvent evaporation and dissolution-precipitation methods [3]. In the present work, commercially available FAHSs (Jingsheng Minerals Co., Ltd., China) were used as raw material, which were calcined at 900°C for 0.5 h in air before use to ensure the mechanical strength. Table 2.1 shows the element compositions of the calcined FAHSs, which mainly consists of SiO_2 and Al_2O_3. The other ingredients in FAHSs such as Fe, K, Ca and Ti would provide a liquid phase during heat treatments promoting particle rearrangement and mass transfer [4]. Polyamide12 powders (PA12, Xincheng Engineering Plastics Co., Ltd., China) with an average particle size of 55.0 μm and apparent density of 0.48 g/cm^3 were used as the binders to mechanical mixed FAHSs. Meanwhile, the PA12 pellets (Degussa Co., Germany) with a density of 1.01 g/cm^3 were used to prepare uniform PA12 shell on surface of FAHSs via dissolution-precipitation method, also known as thermally induced phase separation described in the investigation by Zhu et al. [5]. The Al_2O_3 powder (GT3000SG, Almatis, Ludwigshafen,

Germany) with a purity of 99% and average particle size of around 0.3 µm was used as a powder bed, which provides good support for samples during binder removal process to prevent the deformation or cracks.

Table 2.1 Element compositions of raw FAHSs.

Element	Si	Al	Fe	K	Ti	Ca	Na	Mg	Others
Content (wt.%)	52.64	30.88	5.83	3.91	1.94	1.64	0.78	0.73	1.65

2.1 FAHSs/PA12 composites preparation for SLS

In this study, the FAHSs/PA12 composite powders for SLS were prepared by mechanical mixing and dissolution-precipitation method, respectively. As for mechanical mixing method (shown in Figure 2.1), the FAHSs mixed with PA12 directly with different mass ratio through a three-dimensional blender. The calcined FAHSs are in a typical spherical shape with hollow cavity, which possess good flowability suitable for powder deposition (shown in Figure 2.1(b)). After mechanical mixing as shown in Figure 2.1(c), the mixed composite powders are uniform and the FAHSs remain spherical shapes, which exhibit excellent flowability and formability appropriate for SLS. The FAHSs/PA12 composite powders prepared by mechanical mixing method have the advantages of simple operation, short cycle time and low cost. However, the uniformity of composite powders is easily affected by external factors such as the particle size of binders and rotating speed of blender, which probably causes the powder segregation during laser deposition.

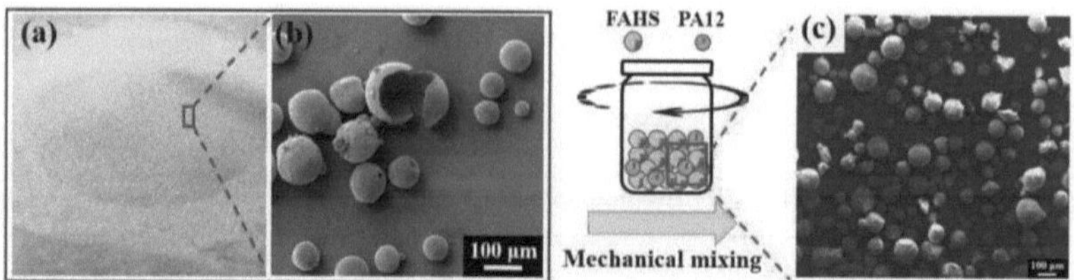

Figure 2.1 Schematic diagram of preparation of ceramic foams by SLS: (a) photograph and (b) SEM of calcined FAHSs; (c) SEM of FAHSs/PA12 composites with 15wt% PA12 [6]

It is reported that the polymer coated ceramic powders, compared with the mechanical mixed powders, are not prone to form segregation during powder deposition and layer sintering due to the relative good flowability and uniform composition. And the prepared green samples exhibit smaller deformation and more uniform internal components [7]. Therefore, in the present work, the core-shell structured PA12/FAHSs composites suitable for SLS were prepared by dissolution-precipitation method using the self-built high temperature reaction kettle, as shown in Figure 2.2, which mainly consists of the heating/cooling circulation pump, panel controller (PC), temperature and stirring speed controller, stirring rotor, exhaust valve and the reactor. The preparation schematic diagram of the core-shell structured PA12/FAHSs composites for SLS experiments is shown in Figure 2.3(a). Firstly, the FAHSs and PA12 pellets with different mass ratio were added into anhydrous ethanol in a reaction kettle and stirred vigorously with a rate of 400 r/min. Then the mixture was heated to

145°C with a heating rate of 5°C/min and holding for 2 h, at which the PA12 pellets dissolved thoroughly to form a homogeneous nylon solution. Subsequently, the solution was gradually cooled down to 107°C holding for 1 h, at which the PA12 crystallized taking the FAHSs as heterogeneous nuclei. When the mixture was cooled down to room temperature, the core-shell structured PA12/FAHSs composites were obtained after filtration, drying, and sieving.

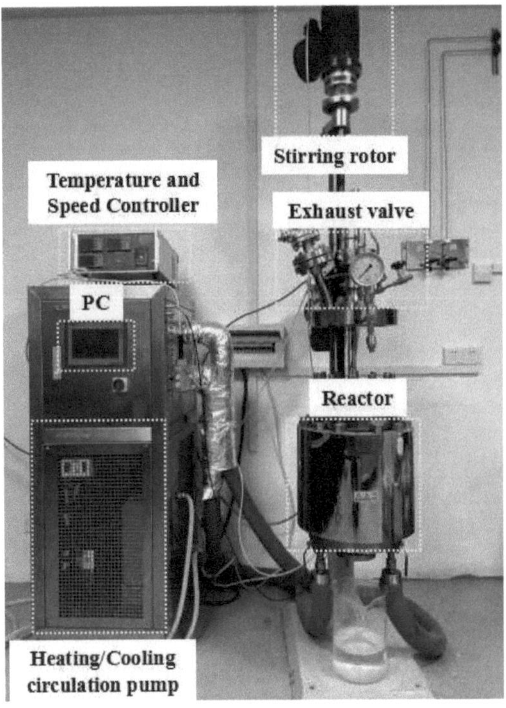

Figure 2.2 High-temperature reaction kettle used for the preparation of core-shell structured PA12/FAHSs composites by dissolution-precipitation method

Figure 2.3(c-f) show the microstructure of FAHSs coated with different PA12 contents via dissolution-precipitation method. All PA12/FAHSs composites exhibit a spherical shape, which show good flowability that allows homogeneous powder deposition for SLS. The surfaces of composites are coarse due to the generation of PA12 shell with a thickness of several microns, and therefore as shown in Figure 2.3(g), the composites show a relatively larger average diameter compared with the calcined FAHSs (shown in Figure 2.3(b)).

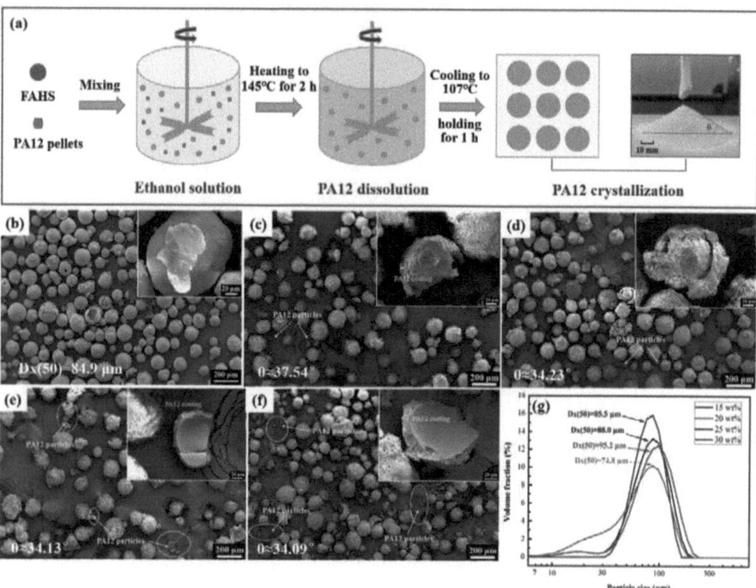

Figure 2.3. (a) Preparation schematic diagram of FAHSs coated with PA12 shell for SLS experiments, SEM images of (b) FAHSs calcined at 900°C for 0.5 h and then coated with different PA12 contents: (c) 15 wt%, (d) 20 wt%, (e) 25 wt% and (f) 30 wt%, and (g) the average particle size distribution of the composite powders [8]

It should be noted that with the increase of PA12 content, the average particle size distribution of core-shell structured PA12/FAHSs composites firstly increases from 88.0 µm to 95.2 µm and then decreases to 74.8 µm which is even smaller than that of calcined FAHSs. Such phenomenon probably results from the self-crystallization and nucleation of PA12 during the dissolution-precipitation method. As shown in Figure 2.4(a), when PA12 content is below 20 wt%, it crystallizes taking the FAHSs as heterogeneous nuclei preferentially as the energy required for crystallization under the circumstances is relatively low [5]. With the further increase of PA12 content (shown in Figure 2.4(b)), however, part of PA12 would self-crystallize and nucleate since the number of FAHSs required for heterogeneous nucleation is limited, and in this case, more self-nucleated PA12 powders with relatively small particle size would be generated as shown in Figure 2.3(e-f). Therefore, the average particle size distribution of PA12/FAHSs composites drops gradually with the further increase of PA12 content from 20 wt% to 30 wt%. Meanwhile, the flowability of the composites is improved since the free flow angle (θ) (shown in Figure 2.3(a)) decreases from 37.54° to 34.09° with increasing PA12 content. However, excessive binder content might result in a large shrinkage after high temperature sintering, which causes the deformation and cracking easily.

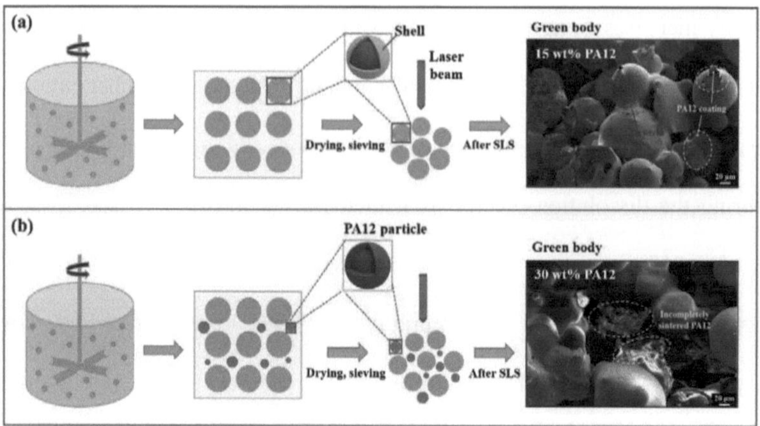

Figure 2.4 Scheme diagram of the SLS fabricated ceramic green body using FAHSs
with PA12 addition (a) below and (b) above 20 wt% [8]

2.2 Ceramic foam preparation by SLS

The ceramic green bodies were prepared through SLS experiments
performed on the C250 SLS machine (Wuhan Huake 3D Technology Co.
Ltd., China) equipped with CO_2 laser beam, as shown in Figure 2.5.

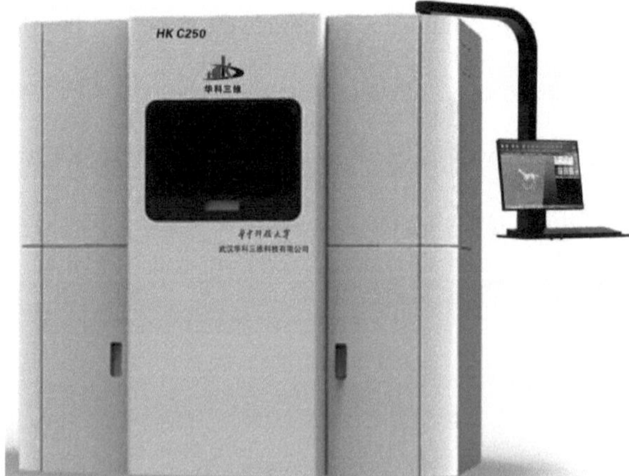

Figure 2.5 C250 SLS machine used for SLS experiments

The photograph of powder bed for layer deposition in SLS machine is shown in Figure 2.6, which mainly consists of the feed bed, build bed and the roller system. The device parameters of C250 SLS machine equipped with CO_2 laser is shown in Table 2.2.

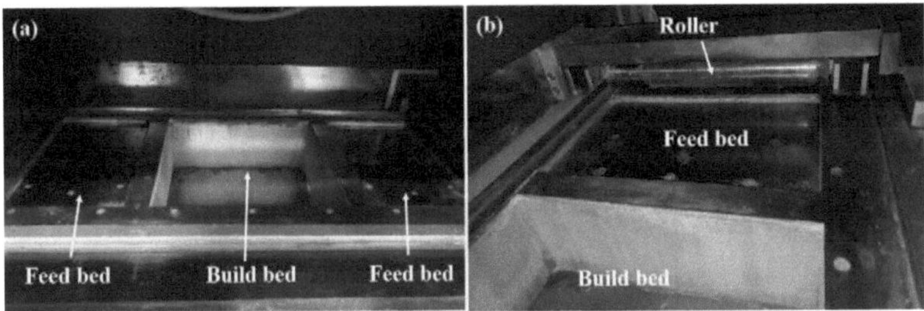

Figure 2.6 Photograph of powder deposition bed in C250 SLS machine

Table 2.2 Device parameters of C250 SLS machine equipped with CO_2 laser

Model	HK C250
Nominal output power	55 W
Laser spot size	≤ 0.4 mm
Maximum scan rate	5 m/s
Layer thickness	0.08-0.2 mm
Forming precision	100 ± 0.2 mm (0.2%)
Dimensions of forming chamber	250×250×250 (mm)
Controlling software	HUST 3DP (independent R&D)
Physical dimensions (WxHxD)	1650×900×1800 (mm)

Summary

i. The flowbility of mechanical mixed FHASs/PA12 composites is appropriate for laser deposition during SLS.

ii.   Core-shell structured PA12/FAHSs composites prepared by dissolution-precipitation method shows a better flowability and formability suitable for SLS.

iii.   The free flow angle of core-shell structured composites decreases from $37.54°$ to $34.09°$ with an increasing PA12 content, which indicates the improved flowability.

iv.   The average particle size of core-shell structured PA12/FAHSs composites first increases from 88.0 μm to 95.2 μm and then decreases to 74.8 μm with increasing the content.

References

[1] Travitzky N., Bonet A., Dermeik B., et al. Additive manufacturing of ceramic‐based materials. Advanced Engineering Materials, 2014, 16(6): 729-754.

[2] Liu K., Shi Y., He W., et al. Densification of alumina components via indirect selective laser sintering combined with isostatic pressing. The International Journal of Advanced Manufacturing Technology, 2013, 67(9-12): 2511-2519.

[3] Xia S., Liu K. Study on preparation, SLS and post-process of ZrO coated powder. Proc. 1st Int. Conf. on 'Progress in Additive Manufacturing', Singapore, May 2014.

[4] Mollah M.Y.A, Promreuk S., Schennach R., et al. Cristobalite

formation from thermal treatment of Texas lignite fly ash. Fuel, 1999, 78(11): 1277-1282.

[5] Zhu W., Yan C., Shi Y., et al. A novel method based on selective laser sintering for preparing high-performance carbon fibres/polyamide12/epoxy ternary composites. Scientific reports, 2016, 6: 33780.

[6] Li M., Chen A.N., Lin X., et al. Lightweight mullite ceramics with controlled porosity and enhanced properties prepared by SLS using mechanical mixed FAHSs/polyamide12 composites. Ceramics International, 2019, 45(16): 20803-20809.

[7] Xiong B.W., Yu H., Xu Z.F., et al. Study on dual binders for fabricating SiC particulate preforms using selective laser sintering. Composites Part B: Engineering, 2013, 48: 129-133.

[8] Chen A.N., Gao F., Li M., et al. Mullite foams with controlled pore structures and low thermal conductivity prepared by SLS using core-shell structured polyamide12/FAHSs composites. Ceramics International, 2019, 45(12): 15538-15546.

3 Characterization

3.1 Characterization of powder and green bodies

The diameter distribution of raw FAHSs and spherical PA12 powders was evaluated using a laser diffraction particle size analyzer (Mastersizer 3000, Worcestershire, United Kingdom). The element composition was characterized by the X-ray Fluorescence Spectrometer (Zetium, PANalytical, Holland). The apparent density of FAHS powders was measured as 0.43 g/cm^{-3} using a powder comprehensive characteristics tester (BT-1000, Bettersize Instruments Ltd., China). Thermo-gravimetric analysis (TGA) of the green body and PA12 powder was measured by a differential scanning calorimeter (DSC, Diamond, PerkinElmer Instruments Inc., Shanghai, China).

3.2 Microstructural and phase characterization of ceramic foams

The morphological and microstructural characterizations of ceramic samples were performed on a scanning electron microscope (SEM, JSM-7600F, JEOL Ltd., Japan) coupled with an energy dispersive spectroscopy (EDS). Transmission electron microscope (TEM) and high resolution transmission electron microscope (HRTEM) measurements, together with EDS analysis (Tecnai G2 F30 S-TWIN, FEI, Holland) were conducted to observe the morphology and chemical compositions of sintering necks in ceramic samples. The crystal structure was investigated by selected area electron diffraction (SAED) based on TEM. Phase compositions of ceramic foams sintered at different temperatures were characterized by X-ray diffractometer (XRD-7000s, Shimadzu, Japan) with a Cu tube (λ=1.5406 Å) at 40 kV and 30 mA. The scattering angular (2θ) varied from 10-80° with a scan rate of 5°/min.

3.3 Pore structure characterization of ceramic foams

The closed pore size distribution of mullite foams was evaluated from more than 400 counts based on SEM images using Nano Measurer software, which has been applied in many other researches [1-3]. The open pore size distribution and open porosity of ceramic foams were measured by a high-performance automatic mercury pressure meter (AutoPore IV 9500, Micromeritics Instrument Corp., Shanghai, China) under the pressure from 0 to 30,000 psia. The geometric density of FAHS ceramic foams was measured by calculating the ratio of mass to volume. The true density of raw FAHSs was evaluated by a helium pycnometer (Micromeritics AccuPyc 1330, Norcross, GA). The total porosity (θ) of ceramic foams was calculated from equation (3.1):

$$\theta = \frac{1-\rho_1}{\rho_0} \times 100\% \qquad\qquad (3.1)$$

where ρ_1 and ρ_0 are the bulk density and true density of FAHS ceramic foams, respectively.

3.4 Mechanical property characterization of ceramic foams

The shrinkage of the prepared ceramic foams was calculated according to equation (3.2):

$$S = \frac{l_0 - l}{l_0} \times 100\% \tag{3.2}$$

where S is the linear shrinkage, l_0 and l are the diameter of the cylindrical green body and ceramic sample, respectively. Cylindrical ceramic bodies with around 15.0 mm in diameter and a height between 15 and 18 mm were prepared for compressive strength measurements, which were performed on a mechanical testing machine (AG-100KN, Zwick/Roell, Germany) with a crosshead loading speed of 0.5 mm/min. The illustrated data represents the average value from 5 samples.

3.5 Thermal conductivity characterization of ceramic foams

The thermal diffusion coefficient of the ceramic foams was measured using the laser thermal conductivity analyzer (LFA 427, NETZSCH-Gerätebau GmbH, Selb, Germany) with laser voltage of 450 V and pulse width of 0.50 milliseconds. The thermal conductivity (λ) of sintered samples was calculated from Equation (3.3) [4].

$$\lambda = \rho \times \alpha \times c_m \tag{3.3}$$

where ρ is the theoretical density of FAHSs, α and c_m are the thermal diffusion coefficient and specific heat capacity of ceramic foams, respectively.

References

[1] Wei X.Q., Payne G.F., Shi X.W., et al. Electrodeposition of a biopolymeric hydrogel in track-etched micropores. Soft Matter, 2013, 9(7): 2131-2135.

[2] Jiang C., He H., Jiang H., et al. Nano-lignin filled natural rubber composites: Preparation and characterization. Express Polymer Letters, 2013, 7(5).

[3] Chen A.N., Wu J.M., Han L.X., et al. Preparation of Si_3N_4 foams by DCC method via dispersant reaction combined with protein-gelling. Journal of Alloys and Compounds, 2018, 745: 262-270.

[4] Wu J., Wei X., Padture N.P., et al. Low thermal conductivity rare earth zirconates for potential thermal barrier coating applications. Journal of the American Ceramic Society, 2002, 85(12): 3031-3035.

4 Laser sintering and high-temperature sintering

In SLS technology, the forming quality of ceramic green bodies is related not only to the inherent characteristics of powder materials, but also to the processing parameters such as laser power, scanning speed, hatch spacing, layer thickness, preheating temperature and so on, which has been discussed in detail in previous research works [1-4]. Ho et al. has demonstrated that one of the key processing parameters for SLS technology is the laser energy density, which can be expressed as following Equation [5]

$$e = \frac{P}{H \times v} \tag{5}$$

where e is the laser energy density, P is the laser power, H is the hatch spacing and v is the scanning speed. It can be seen from Eq. (5) that the laser power has a great influence on the forming quality of green bodies, which determines the energy output of the laser beam. A high level laser energy density would be generated from the high laser power, causing the burning or evaporation of polymer binders losing the bonding effect. On the contrary, the poor adhesion of binders could not allow the sufficient bonding of powders when the laser power is not high enough, resulting in the lack of strength of green bodies. Therefore, the applied laser power should be higher than a certain critical value leading to the strong bonding of powder in the scanning area while the powder in surrounding non-scanning area could not be bonded [6]. The laser energy density also increases with decreasing scanning speed at the same level of laser power, which expands the width and depth of the melting zone contributing to strength improvement of green bodies. However, in practical forming process, lower scanning speed would expand the heat-affected zone, which reduces the forming accuracy of products and causes the warping deformation [7]. The hatch spacing refers to the distance between two adjacent scanning lines, which could directly control the energy superposition of the adjacent lines. An ideal hatch spacing in SLS should be able to keep the continuity and integrity of powder bonding in every single layer and avoid the over burnt in overlap scanning field [8]. Typically, an orthogonal experiment is generally carried out to evaluate the formability of powders using different combinations of laser scanning parameters in SLS [9].

4.1 Influence of laser sintering parameters on green bodies

In this study, the laser powers of 6.0, 6.6 and 7.2 W, the scanning speeds of 1600, 1800 and 2000 mm/s, and the hatch spacing of 110, 130 and 150 μm are selected for the orthogonal experimental test (shown in Table 4.1). Each parameter combination with layer thickness of 130 μm is employed to fabricate the ceramic green body. The effect of laser scanning parameters on the shrinkage and compressive strength of green samples prepared using 15 wt% PA12 addition is shown in Figure 4.1. The shrinkage of green samples increases as increasing the laser power, while it shows the opposite tendency with the increase of scanning speed and hatch spacing. Furthermore, the compressive strength of green bodies increases sharply with increasing laser power, while the scanning speed and hatch spacing have little impact on the strength change. Therefore, the SLS processing parameter combination, i.e., laser power of 6.6 W, scanning speed of 1800 mm/s and hatch spacing of 130 µm can be selected for the following SLS experiment as its samples show the relatively low linear shrinkage and high compressive strength.

Table 4.1 Levels of laser scanning parameters of SLS

No.	Parameters	Level 1	Level 2	Level 3
1	Laser power (W)	6.0	6.6	7.2
2	Scanning speed (mm/s)	1600	1800	2000
3	hatch space (µm)	110	130	150

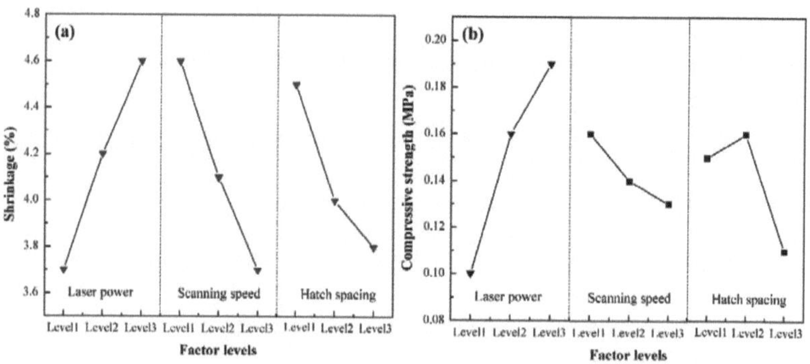

Figure 4.1 (a) Shrinkage and (b) compressive strength of green bodies prepared using 15 wt% PA12 content under different laser parameter combinations [10]

The honeycomb ceramic green body with intersecting holes is successfully prepared by SLS using the parameter combination (laser power of 6.6 W, scanning speed of 1800 mm/s and hatch spacing of 130

μm). As shown in Figure 4.2, no obvious delamination or cracking can be

observed in the fabricated green body.

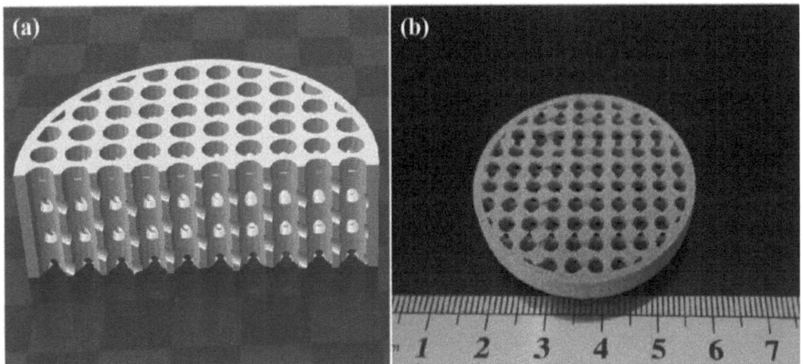

Figure 4.2 (A) honeycomb 3D model and (b) ceramic green body prepared by SLS

4.2 Determination of high temperature sintering process

Figure 4.3 shows the TGA curves of PA12 powder and green body

prepared by SLS. The PA12 powder starts evaporating at 330°C and loses

about 93 wt% of its weight at 650°C. The green body starts evaporating at

about 330°C and a mass loss of 20% occurs before 650°C mainly due to

the burnt out of PA12 binder phase, which could melt and bond FAHSs

during laser irradiation. Therefore, the heating cycle of the green body

contains two steps based on the TGA data. The first is heating to 650°C for

2 h with a heating rate of 1°C/min to remove the binder phase. Then the

temperature is increased to the expected sintering point (1250°C-1400°C)

with a heating rate of 5°C/min and holding for 3 h to obtain the final

ceramic foams.

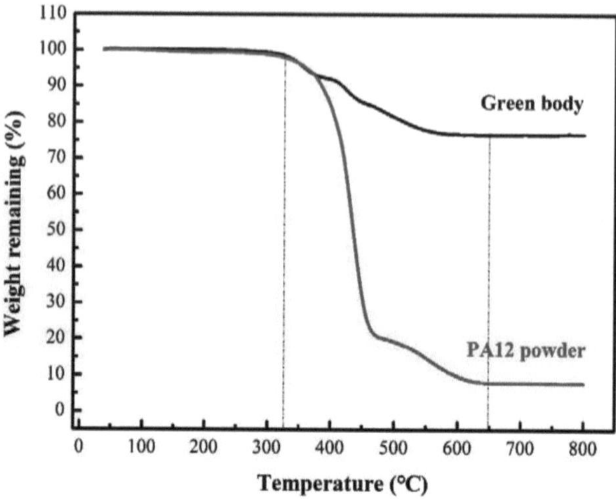

Figure 4.3 TGA curves of PA12 powder and the prepared green body by SLS

Summary

i.   SLS processing parameters, i.e., laser power of 6.6 W, scanning speed of 1800 mm/s and hatch spacing of 130 μm are selected for experiment.

ii.   Crack-free honeycomb ceramic green body with intersecting holes is successfully prepared by SLS.

iii.   Green bodies were firstly calcined at 650°C for 2 h with a heating rate of 1°C/min, and then sintered at 1250-1400°C and holding for 3 h with a heating rate of 5°C/min.

Reference

[1] Liu K., Shi Y., He W., et al. Densification of alumina components via indirect selective laser sintering combined with isostatic pressing. The

International Journal of Advanced Manufacturing Technology, 2013, 67(9-12): 2511-2519.

[2] Shahzad K., Deckers J., Kruth J.P., et al. Additive manufacturing of alumina parts by indirect selective laser sintering and post processing. Journal of Materials Processing Technology, 2013, 213(9): 1484-1494.

[3] Deckers J., Meyers S., Kruth J.P., et al. Direct selective laser sintering/melting of high density alumina powder layers at elevated temperatures. Physics Procedia, 2014, 56: 117-124.

[4] Shahzad K., Deckers J., Zhang Z., et al. Additive manufacturing of zirconia parts by indirect selective laser sintering. Journal of the European Ceramic Society, 2014, 34(1): 81-89.

[5] Ho H.C.H., Gibson I., Cheung W.L., Effects of energy density on morphology and properties of selective laser sintered polycarbonate. Journal of Materials Processing Technology, 1999, 89: 204-210.

[6] Xia S., Fabricating silicon carbide parts by selective laser sintering/cold isostatic pressing and post process. Master's dissertation, Huazhong University of Science and Technology, Hubei, China, 2016.

[7] Xu W.,Study on selective laser sintering of SiC and post pocess. Master's dissertation, Huazhong University of Science and Technology, Hubei, China, 2007.

[8] Cheng D.,Study on selective laser sintering of aumina parts and post pocess. Master's dissertation, Huazhong University of Science and

Technology, Hubei, China, 2007.

[9] Chen A.N., Li M., Xu J., et al. High-porosity mullite foams prepared by selective laser sintering using fly ash hollow spheres as raw materials. Journal of the European Ceramic Society, 2018, 38(13): 4553-4559.

[10]　Chen A.N., Gao F., Li M., et al. Mullite foams with controlled pore structures and low thermal conductivity prepared by SLS using core-shell structured polyamide12/FAHSs composites. Ceramics International, 2019, 45(12): 15538-15546.

5 Properties of SLS-formed mullite foams

5.1 Microstructure and phase analysis

After high temperature sintering at 1250-1400°C, the XRD patterns of ceramic foams are shown in Figure 5.1. When sintered at 1250-1350°C, the peak of cristobalite phase can be detected which derives from the amorphous glassy phase. The cristobalite phase develops through the liquid phase provided by alkalies and alkaline earth elements during heat treatments, as reported in [1]. With increasing sintering temperature, the relative peak intensity of cristobalite phase decreases gradually and disappears at 1400°C as it transforms into amorphous phase again [2,3]. The peak of mullite phase sintered at 1300-1400°C shifts slightly to the left compared with 1250°C, which indicates an expansion of lattice parameters of mullite grains due to the dissolving of larger atoms into the matrix.

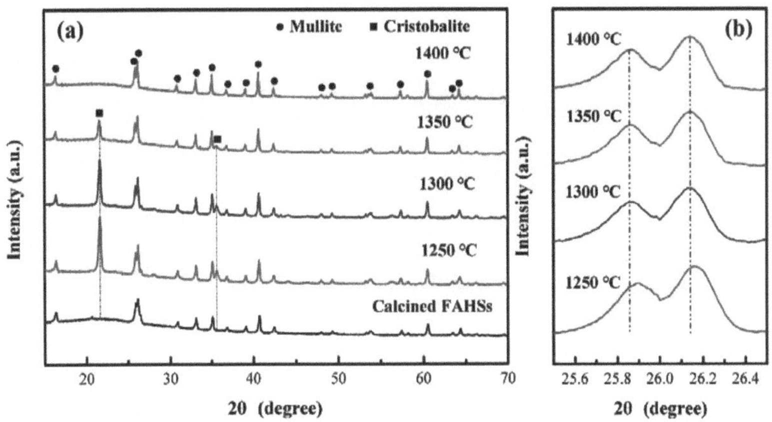

Figure 5.1 XRD patterns of the calcined powder and FAHS ceramic foams sintered at different temperatures [4]

A bimodal pore structure could be observed in mullite foams, i.e., closed pores from enclosed hollow spaces and open pores among FAHSs (shown in Figure 5.2). The bulk foams are constructed by bonding individual hollow spheres together through the sintering necks generated under elevated temperature. With increasing sintering temperature from 1250°C to 1400°C, the sintering necks between FAHSs are promoted. Consequently, the morphology of ceramic foams sintered below 1350°C exhibits intact spheres in natural fracture surface, while that of foams sintered above 1350°C exhibits fractured spheres. Therefore, the fracture mechanism changes from fracturing along FAHSs to across FAHSs as increasing sintering temperature. It can be speculated that the sintering necks generated at elevated temperature might play a key role in mechanical performance of ceramic foams. Figure 5.2(e-f) shows the EDS results over regions of sintering neck (marked as A) and shell wall (marked as B) sintered at 1350°C. In addition to a large amount of Al and Si, some metal elements including 21.15 wt% Fe, 5.50 wt% Ti, 5.28 wt% K and 5.09 wt% Ca can be detected in the sintering neck (shown in Figure 5.2(e)). However, fewer number and content of metal elements, i.e., 4.2 wt% Fe, 1.40 wt% Ti and 1.39 wt% K, are detected in the sphere shell wall (shown in Figure 5.2(f)). The presence of the metal elements provides a liquid phase during heat treatment, which flows into the contact regions between FAHSs to reinforce the sintering necks, corresponding to other studies [5,6].

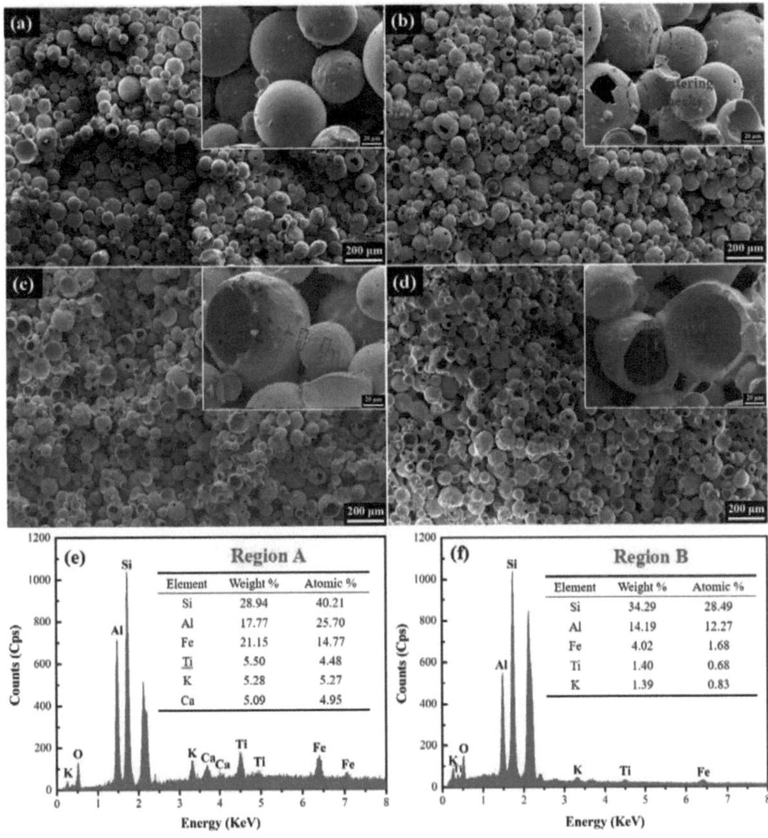

Figure 5.2 SEM images of FAHS ceramic foams sintered at different temperatures (a) 1250°C, (b) 1300°C, (c) 1350°C, (d) 1400°C, and EDS results over several regions (e) sintering neck and (f) FAHS shell of sintered samples at 1350°C [4]

Figure 5.3 shows the morphology of mullite grains on FAHS shell sintered at different temperatures. It can be observed that large amount of elongated mullite grains are formed when sintered at 1250°C and the grains grow gradually with increasing sintering temperature to 1400°C, resulting in the increase of mullite peak intensity as shown in Figure 5.1(a).

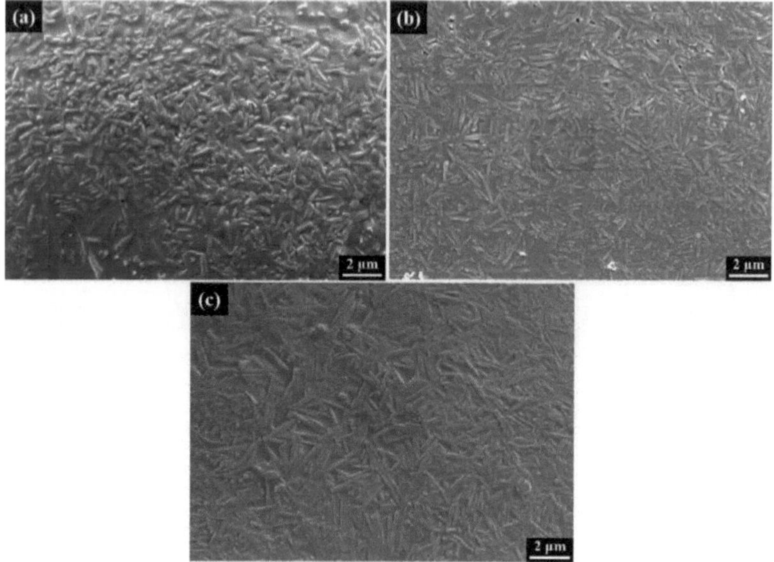

Figure 5.3 Morphology of mullite grains on FAHS shell sintered at different temperatures (a) 1250°C, (b) 1350°C and (c) 1400°C [7]

To further investigate the characteristics of sintering necks between spheres in mullite ceramics, the TEM micrograph together with EDS analysis of sintering neck is shown in Figure 5.4. The mullite grains are in the typical rod shapes with a broad size distribution about 0.2-0.6 μm in width and 0.3-1.7 μm in length randomly embedded in the amorphous glassy phase (shown in Figure 5.4(a)). The EDS elemental mapping images over the blue box region in sintering neck is shown in Figure 5.4(b). The rod-shaped mullite grains are mainly composed of Al and Si, and at the same time Fe is basically dissolved into the mullite grains. However, as for the glassy phase, in addition to large amount of Si, metal elements including K, Ca and Ti can also be detected. Therefore, the reinforcement of sintering necks at high temperature is mainly attributed to the presence of K, Ca and Ti elements in fly ash, which diffuse as the liquid phase into sintering necks to reinforce the bonding. Figure 5.4(d) shows the SAED pattern of the mullite crystal that is marked by "M1" in Figure 5.4(c), showing an orthorhombic structure in [00-1] zone axis. Figure 5.4(f) shows the HRTEM image of yellow dotted box region in M1. The mullite lattice parameters of a, b and c axis determined from the SAED pattern are 3.634 Å, 3.543 Å and 2.554 Å, respectively. They are a little larger than the lattice parameters $a0$ (3.430 Å), $b0$ (3.401 Å) and $c0$ (2.429 Å), which might be attributed to the dissolution of Fe atom into mullite grain (also shown in Figure 5.4(b)), distorting the mullite lattice and expanding the lattice parameters, in agreement with XRD analysis in Figure 5.1(b).

For porous mullite ceramics prepared by SLS using PA12/FAHSs composites, the PA12 binder content would have a great influence on microstructure and properties of the final ceramic products. Figure 5.5 plots the XRD patterns of porous mullite ceramics sintered at 1350°C with different PA12 contents. The peak of mullite and cristobalite phases can be detected in all sintered ceramic samples. XRD results in Figure 5.5(b) clearly indicate that with the increase of PA12 content from 15 wt% to 30 wt%, the intensity of characteristic peak (101) of the cristobalite phase increases gradually suggesting an increasing phase content.

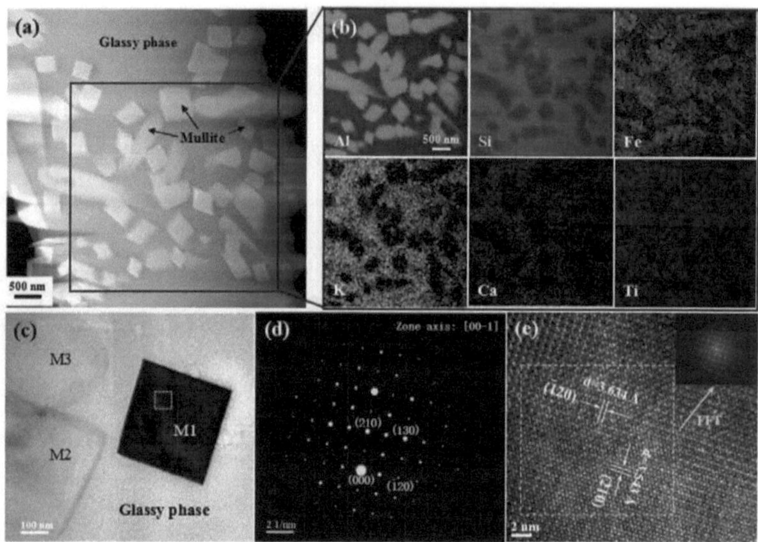

Figure 5.4 (a) Low-magnification TEM micrograph of mullite sintered at 1350°C and (b) their EDS mappings over blue box region. (c) Higher-magnification image of mullite grains, and the (d) SAED patterns and (e) HRTEM image [4]

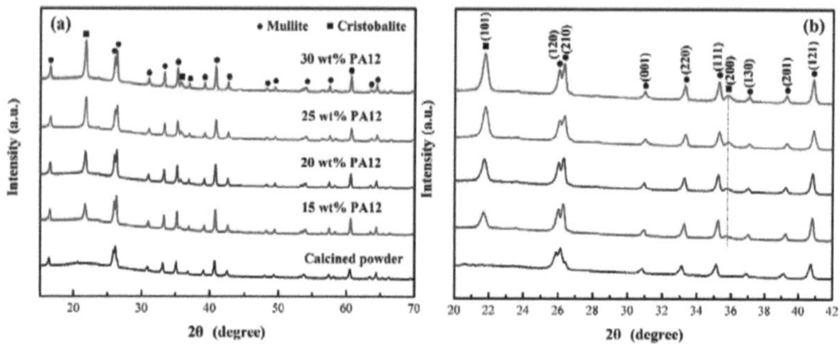

Figure 5.5 (a-b) XRD patterns of porous mullite ceramics sintered at 1350°C with different PA12 contents [8]

According to the above XRD patterns, the mullite main peak ratio can be calculated according to Equation (5.1),

$$M\% = \frac{[I_{M(210)}+I_{M(121)}]}{[I_{M(210)}+I_{M(121)}+I_{C(101)}+I_{C(200)}]} \times 100\% \tag{5.1}$$

where $I_{M(210)}$, $I_{M(121)}$ and $I_{C(101)}$, $I_{C(200)}$ represent the intensities of characteristic peak of mullite and cristobalite phase, respectively. The main peak ratio of mullite phase plotted for ceramic foams is shown in Figure 5.6. This peak ration could not exactly reveal the quantitative ratio, however, it can provide approximate semi-quantitative information about phase content [36]. It is observed that the amount of mullite phase decreases with an increasing PA12 content. Such a phase reduction may result from the reduced particle rearrangement and mass transfer rate as the clearance between solid grains increases with raising PA12 content.

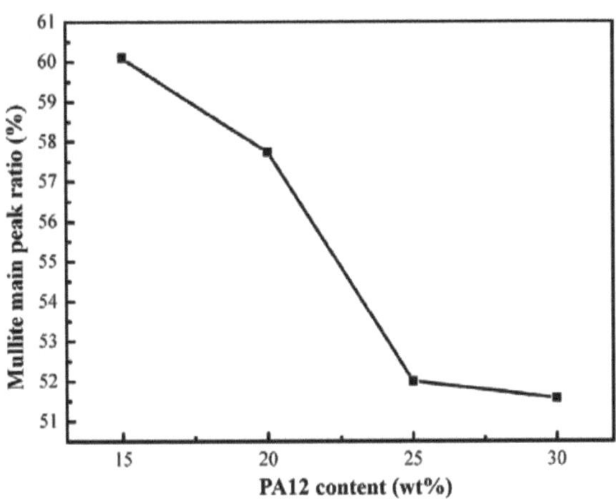

Figure 5.6 Mullite main peak intensity ratio with a variation of PA12 contents [8]

5.2 Pore structure analysis

Since the mullite foams are prepared by SLS only using FAHSs as raw materials, the bulk foams form two kinds of pore structures, i.e., closed and open pores. The closed pore size is determined by the enclosed hollow spaces of FAHSs, while the open pore size is mainly affected by the interspaces among FAHSs. Figure 5.7 shows the closed pore size distribution of FAHS ceramic foams sintered at different temperatures, which is evaluated using Nano Measurer software based on SEM images in Figure 5.2(a-d). With increasing sintering temperature from 1250°C to 1400°C, the average closed pore size gradually decreases from 58.7 μm to 46.0 μm, and the pore size distribution range is getting narrower. This closed pore size drop is mainly caused by the increasing shrinkage of individual spheres as raising the temperature. Therefore, the closed pore size and distribution of FAHS ceramic foams could be well adjusted not only by the particle size and distribution of raw FAHSs but the sintering temperatures. Figure 5.8 shows the open pore size distribution of porous mullite ceramics sintered at different temperatures. Each curve shows a single peak with a narrow width which indicates the uniform open pore size distribution. With increasing the sintering temperature, the open pore size distribution decreases from 48.0 μm to 30.1 μm. This reduction is mainly attributed to two factors. On the one hand, the open pore channels among FAHSs decrease gradually due to the increased shrinkage. On the other hand, the extinction of K, Ca and Ti elements provide a liquid phase under elevated temperature filling the clearance between solid grains, which accelerates particle rearrangement and promotes the densification.

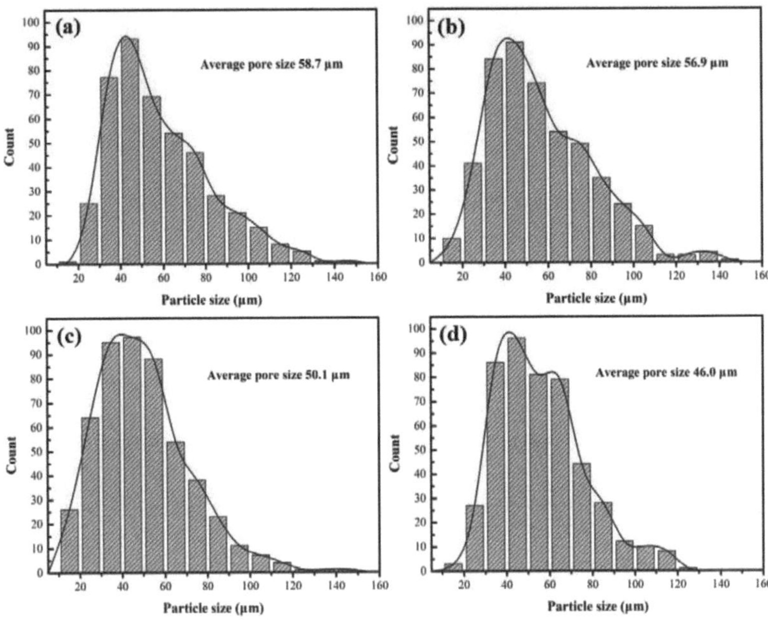

Figure 5.7 Closed pore size distribution of FAHS ceramic foams sintered at different temperatures (a) 1250°C, (b) 1300°C, (c) 1350°C and (d) 1400°C [4]

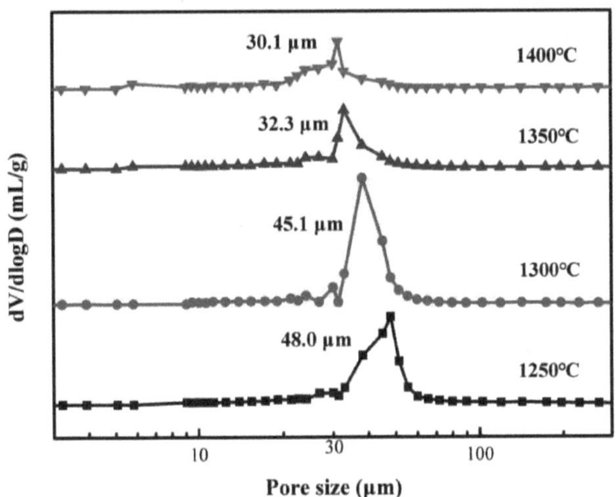

Figure 5.8 Open pore size distribution of FAHS ceramic foams sintered at different temperatures [8]

Figure 5.9 shows the open pore size distribution of FAHS ceramic foams with different PA12 contents sintered at 1350°C. All curves of open pore size distribution show a single peak with a narrow width and the pore size distribution firstly increases from 32.3 μm to 38.3 μm with increasing PA12 content from 15 wt% to 20 wt%, and then keep constant with further increase to 30 wt%. The open pore channels between FAHSs increase in the beginning due to the increased PA12 volume, and when the stacking volume of FAHS reaches the maximum, the excess PA12 will fill the limited gap between the FAHS causing the constant open pore channel size.

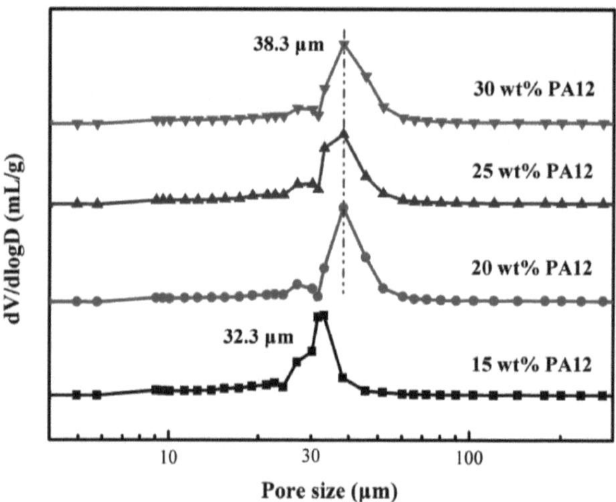

Figure 5.9 Open pore size distribution of FAHS ceramic foams with different PA12 contents sintered at 1350°C [8]

5.3 Mechanical property analysis

It has been proven that the increased sintering temperature could promote the densification. Therefore, the linear shrinkage and bulk density of porous mullites increases from 2.5±0.3% to 13.3±0.7% and from 0.43±0.03 g/cm^3 to 0.43±0.03 g/cm^3, respectively (shown in Figure 5.10). This shrinkage value is much lower than that of ceramic foams prepared by other methods, direct foaming method as an illustration, which achieves a linear shrinkage typically ranging from 20 to 40% [2,3]. The calcined FAHSs have a dense and inactive shell surface which leads to the relatively low and homogeneous shrinkage due to the low sintering activity [9,10].

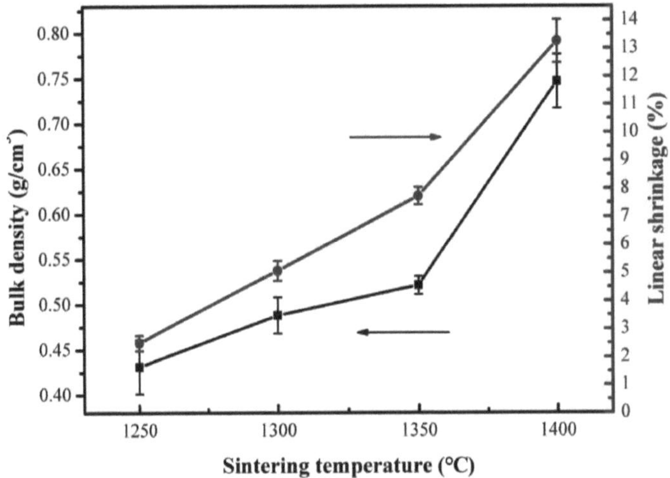

Figure 5.10 Bulk density and linear shrinkage of FAHS ceramic foams sintered at different temperatures [8]

The compressive strength and total porosity of porous mullites sintered at different temperatures are shown in Figure 5.11. As increasing sintering temperature from 1250°C to 1400°C, the compressive strength of porous mullite ceramics increases gradually from 0.3±0.1 MPa to 6.6±0.3

MPa, while the porosity decreases from 86.8±0.5% to 81.0±0.3%. As shown in Figure 5.12, the compressive strength enhancement of FAHS ceramic foams can be divided into two different stages: (I) when sintered at 1250°C, the sintering necks between FAHS are weak and can be easily broken, while the necks are reinforced due to the liquid phase diffusion as increasing sintering temperature, which causes the fracture path change from along sintering necks to across spheres; (II) when raising the temperature above 1300°C, the shell walls are densified and thickened as the shrinkage of sphere increases, and consequently more energy will be required to break the shells. Therefore, the sintering necks between spheres play the key role in strength enhancement at low temperature (<1350°C), while the sphere shell densification acts at high temperature (>1350°C).

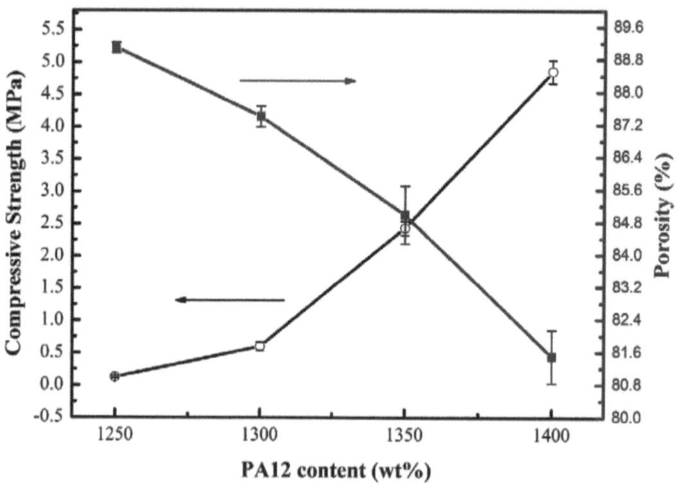

Figure 5.11 Compressive strength and total porosity of porous mullite ceramics sintered at different temperatures [8]

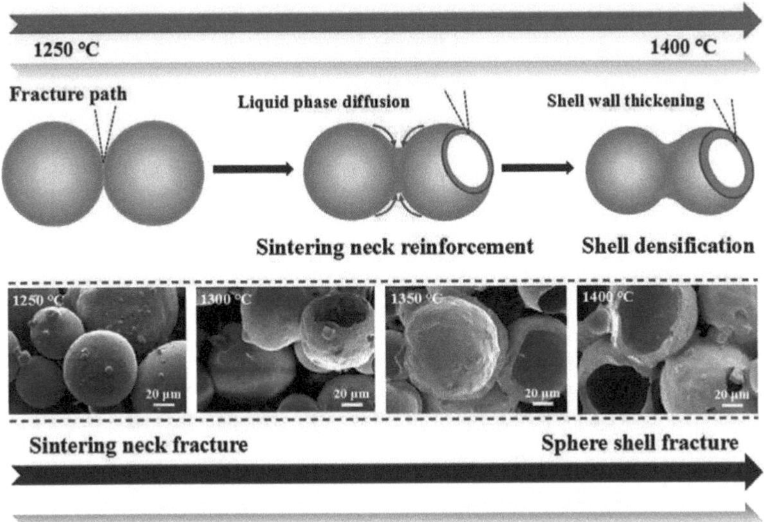

Figure 5.12 Scheme diagram of sintering neck reinforcement and shell densification

of ceramic foams sintered at different temperatures [4]

In addition, it should be noted that the total porosity of ceramic foams

prepared by SLS is much higher than that of foams fabricated by other

conventional methods such as direct stack sintering and gel-casting [3]. As

schematically shown in Figure 5.13(a), the porosity of ceramic foams

prepared by gelcasting or direct stack sintering mainly consists of the inner

hollow structure of spheres (marked as α) and the interspaces among

stacking FAHSs (marked as β). While in the ceramic foams prepared by

SLS, besides the pore structures α and β, the special gaps among FAHSs

directly related to SLS process (marked as γ as shown in Figure 5.13(b))

also exist since the green bodies are porous due to the incompact powder

layers and the sacrificial binder phase.

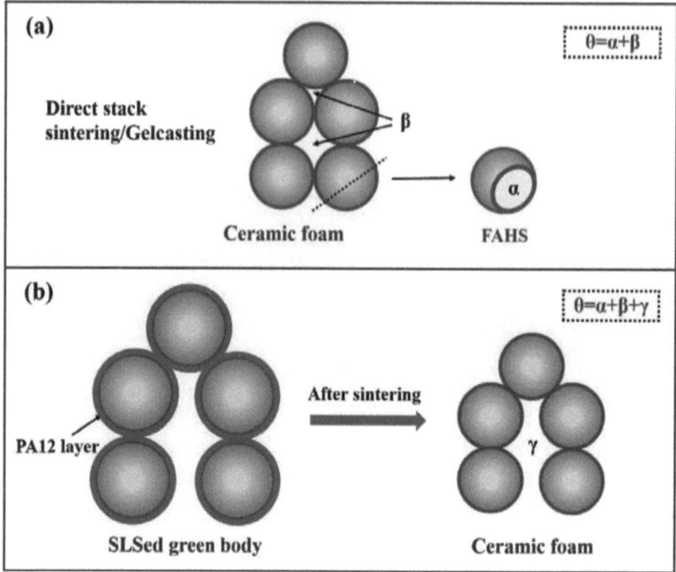

Figure 5.13 Schematic diagram of the pore structure of ceramic foams prepared by (a) direct stacking/gelcasting and (b) SLS using only FAHSs as raw material [8]

The effect of PA12 content on linear shrinkage in different directions of ceramic samples in shown Figure 5.14. It can be observed that the total linear shrinkage of ceramic foams in radial and axial directions basically increases gradually as increasing PA12 content. However, the shrinkage in radial direction is much higher than that in axial direction, which is attributed to the layer-by-layer manufacturing process in SLS [11]. Figure 5.15 shows the compressive strength and total porosity of porous mullite ceramics with different PA12 contents. The total porosity of porous mullite ceramics firstly increases from 85.0±0.5% to 85.4±0.4% and then decreases slightly to 85.0±0.2% with an increasing PA12 content, and the compressive strength correspondingly decreases from 2.4±0.1 MPa to 2.0±0.1 MPa at the initial stage and then slightly increases to 2.61± 0.2MPa.

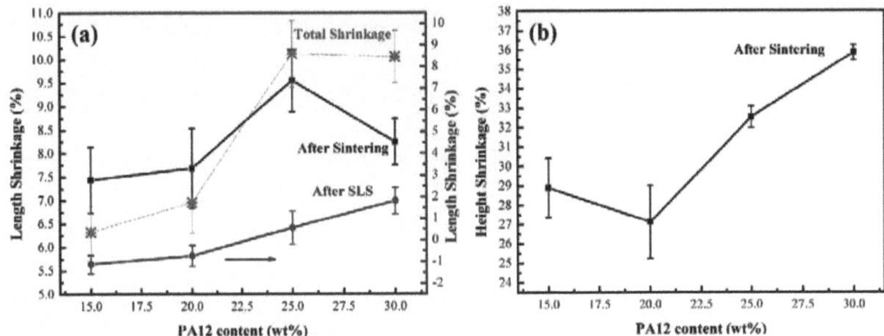

Figure 5.14 Linear shrinkage in radial and axial directions of ceramic samples with different PA12 contents

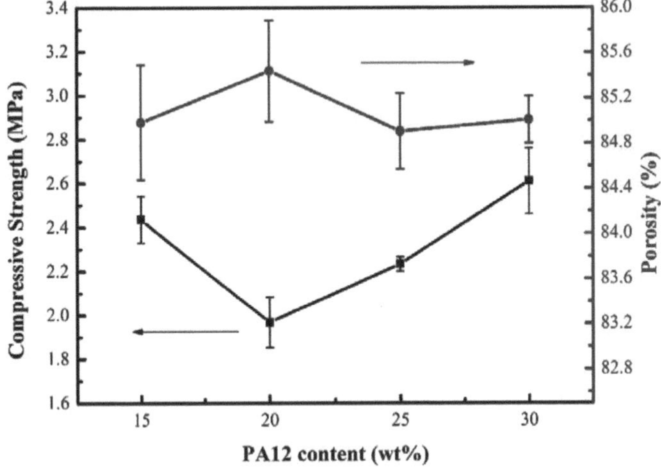

5.15 Compressive strength and total porosity of porous mullite ceramics with different PA12 contents sintered at 1350°C [8]

5.4 Thermal property analysis

Figure 5.16 plots the thermal conductivity of ceramic foams as a function of temperature. The thermal conductivity of all ceramic samples increase gradually with increasing the firing temperature from 25 to 450°C, which could be attributed to the improved heat radiation with the temperature [12]. Furthermore, the thermal conductivity of porous mullite ceramics basically increases with increasing the sintering temperature from 1250°C to 1400°C. As is well known, the thermal conductivity of a given material is generally composed of the electron and the lattice thermal conductivity [13,14]. Generally, the electron thermal conductivity mainly depends on the porosity of materials, while the lattice thermal conductivity depends on the microstructure [15]. In this case, with sintering temperature increasing from 1250°C to 1400°C, the thermal conductivity of porous mullite ceramics increases mainly due to the decreased total porosity for materials with same phase.

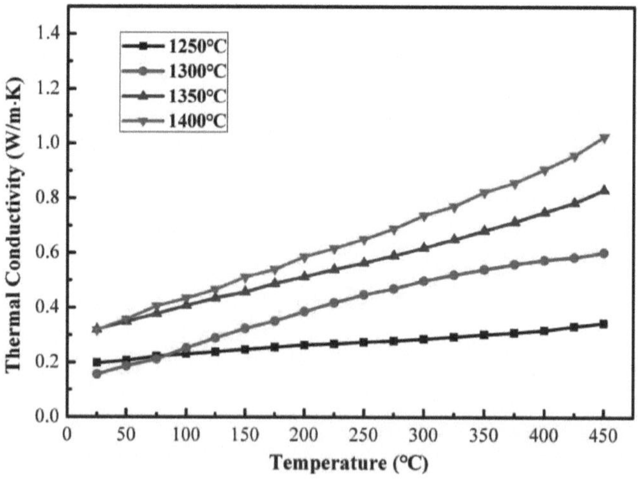

Figure 5.16 Thermal conductivity vs. temperature for porous mullite ceramics sintered at different temperatures [8]

The thermal conductivity of the samples prepared using different PA12 content as a function of temperature is shown in Figure 5.17. The thermal conductivity of all samples increases gradually as increasing the firing temperature from 25 to 450°C, the same trend as the curve shown in Figure 5.16. Moreover, with the increase of PA12 content from 15 w% to 30 wt%, the thermal conductivity of porous mullite ceramics decreases gradually mainly due to the increased cristobalite phase and decreased mullite phase detected in XRD patterns (shown in Figure 5.5), as the thermal conductivity of dense cristobalite and mullite ceramic were chosen to be 2.5 and 5.1 W/(m·K) from literature values [16,17]. Finally, the rather low thermal conductivity of 0.08 W/(m·K) can be obtained when added 30 wt% PA12 content.

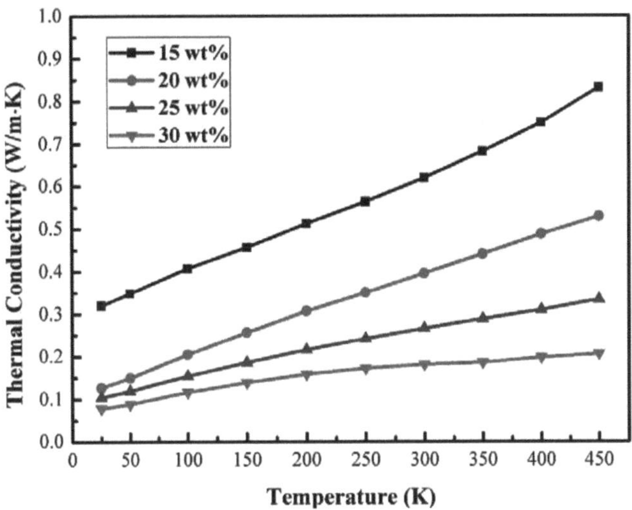

Figure 5.17 Thermal conductivity vs. temperature for porous mullite ceramics prepared with different PA12 contents [8]

In Figure 5.18, we compare the thermal conductivity, porosity and compressive strength of the SLS fabricated highly porous mullite ceramics demonstrated in this work with those of recent publications (data points a-t). Compared with the conventional methods such as dry pressing, gelcasting and gelation freezing, the SLS fabricated porous mullite ceramics in this work exhibit the relatively high porosity and high compressive strength with rather low thermal conductivity, which is probably attributed to the special pore structures of products related to the FAHSs and SLS technique as schematically illustrated in Figure 5.13(b). Most notably the complex shaped mullite ceramic products, for example, the honeycomb ceramics with intersecting holes (shown in Figure 4.2(b)), are rather difficult and challenging to be fabricated by conventional methods, but can be easily implemented by SLS technique. Finally, the crack-free mullite foams with complex shapes are obtained as shown in Figure 5.19. No macroscopic or microscopic cracks can be observed in the foams. The results above suggest that complicated-structured mullite foams could be prepared by SLS using FAHSs as starting material, which is expected to achieve high geometrical complexity and have potential applications in the fields of thermal insulation, heat dissipaters, filter, acoustical insulation, etc [37].

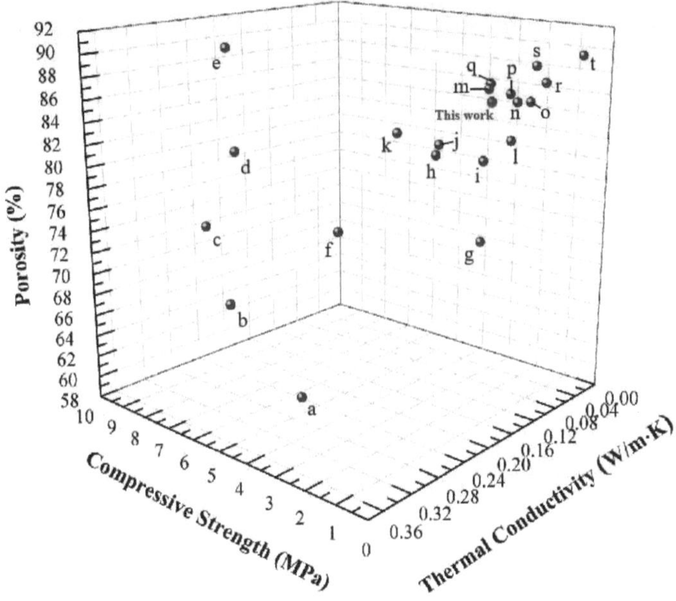

Figure 5.18 Thermal conductivity, porosity and compressive strength of mullite foams (with porosity >60%). The plot summarizes the current performance of the porous mullite ceramics in this work and in prior publication. Points and references include: a) ref. [18], b) ref. [19], c) ref. [20], d) ref. [21], e) ref. [22], f) ref. [23], g) ref. [24], h) ref. [25], i) ref. [26], j) ref. [27], k) ref. [28], l) ref. [29], m) ref. [30], n) ref. [16], o) ref. [31], p) ref. [32], q) ref. [33], r) ref. [34], s) ref. [35], t) ref. [36].

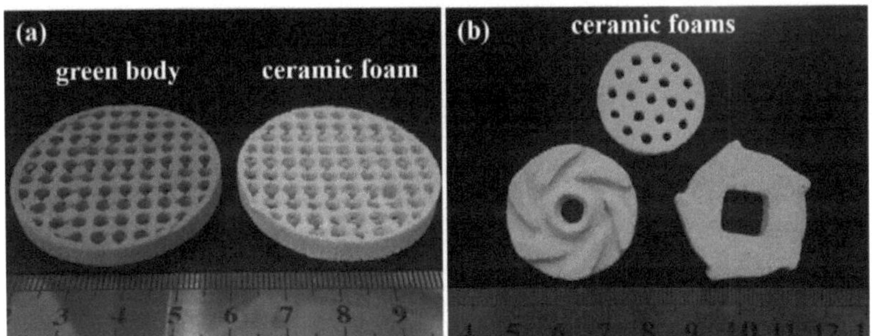

Figure 5.19 Mullite foams with complex shapes prepared by SLS: (a) Honeycomb mullite green body and ceramic product; (b) Other complex-shaped FAHS ceramic foams ceramics

Summary

i. High porosity ranging 79.9-89.1% of ceramic foams is resulted from the inner hollow structure, the interspaces between stacking FAHSs, and the gaps between FAHSs directly related to SLS.

ii. The pore structure features (porosity, closed/open pore size distribution) and mechanical properties of ceramic foams can be well adjusted by the sintering temperature and PA12 content.

iii. The fracture mechanism of FAHS ceramic foams changes from fracturing along spheres to across spheres with increasing sintering temperature.

iv. The strength enhancement of FAHS ceramic foams at low sintering temperature (<1350°C) is attributed to the reinforced sintering necks between spheres, while the densification and thickening of shell walls plays a key role when sintered above 1350°C.

v. Thermal conductivity of 0.08 W/(m·K) can be obtained which is lower than that of ceramics prepared by conventional methods due to the special pore structures in FAHS ceramic foams prepared by SLS.

Reference

[1] Mollah M.Y.A., Promreuk S., Schennach R., et al. Cristobalite formation from thermal treatment of Texas lignite fly ash. Fuel, 1999, 78(11): 1277-1282.

[2] Li N., Zhang X.Y., Qu Y.N., et al. A simple and efficient way to prepare porous mullite matrix ceramics via directly sintering SiO_2-Al_2O_3 microspheres. Journal of the European Ceramic Society, 2016, 36(11): 2807-2812.

[3] Huo W., Zhang X., Chen Y., et al. Novel mullite foams with high porosity and strength using only fly ash hollow spheres as raw material. Journal of the European Ceramic Society, 2018, 38(4): 2035-2042.

[4] Chen A.N., Li M., Wu J.M., et al. Enhancement mechanism of mechanical performance of highly porous mullite ceramics with bimodal pore structures prepared by selective laser sintering. Journal of Alloys and Compounds, 2019, 776: 486-494.

[5] Qi F., Xu X., Xu J., et al. A novel way to prepare hollow sphere ceramics. Journal of the American Ceramic Society, 2014, 97(10): 3341-3347.

[6] Sanders W.S., Gibson L.J. Mechanics of BCC and FCC hollow-sphere foams. Materials Science and Engineering: A, 2003, 352(1-2): 150-161.

[7] Chen A.N., Li M., Xu J., et al. High-porosity mullite foams prepared by selective laser sintering using fly ash hollow spheres as raw materials. Journal of the European Ceramic Society, 2018, 38(13): 4553-4559.

[8] Chen A.N., Gao F., Li M., et al. Mullite foams with controlled pore structures and low thermal conductivity prepared by SLS using core-

shell structured polyamide12/FAHSs composites. Ceramics International, 2019, 45(12): 15538-15546.

[9] Deng Z.Y., Yang J.F., Beppu Y., et al. Effect of agglomeration on mechanical properties of porous zirconia fabricated by partial sintering. Journal of the American Ceramic Society, 2002, 85(8): 1961-1965.

[10] Brasil A.M., Farias T.L., Carvalho M.G., et al. Numerical characterization of the morphology of aggregated particles. Journal of Aerosol Science, 2001, 32(4): 489-508.

[11] Li M., Chen A.N., Lin X., et al. Lightweight mullite ceramics with controlled porosity and enhanced properties prepared by SLS using mechanical mixed FAHSs/polyamide12 composites. Ceramics International, 2019, 45(16): 20803-20809.

[12] Luo X.W., Li H.F., Xiang W.G., et al. Fabrication, properties and nonlinear thermal conductivity model of highly porous mullite ceramics. J. Inorg. Mater, 2014, 29: 1179-1185.

[13] Luo Y., Yang J., Jiang Q., et al. Investigation on the microstructure and thermoelectric performance of magnetic ions doped $Bi_{0.5}Sb_{1.5}Te_3$ solidified under a magnetostatic field. Acta Materialia, 2017, 127: 185-191.

[14] Zhu B.B., Tian R., Zhang T., et al. Tunable thermopower and thermal conductivity in Lu doped In_2O_3. RSC Advances, 2014, 4(60): 31926-31931.

[15] Wang J., Carson J.K., North M.F., et al. A new approach to modelling the effective thermal conductivity of heterogeneous materials. International journal of heat and mass transfer, 2006, 49(17-18): 3075-3083.

[16] Gong L., Wang Y., Cheng X., et al. Thermal conductivity of highly porous mullite materials. International Journal of Heat and Mass Transfer, 2013, 67: 253-259.

[17] Garcıa E., Osendi M.I., Miranzo P. Thermal diffusivity of porous cordierite ceramic burners. Journal of applied physics, 2002, 92(5): 2346-2349.

[18] Xu N.N., Li S.J., Li Y.B., et al. Fabrication and characterisation of porous mullite ceramics from high voltage insulator waste. Advances in Applied Ceramics, 2015, 114(2): 93-98.

[19] Yuan L., Ma B., Zhu Q., et al. Preparation and properties of mullite-bonded porous fibrous mullite ceramics by an epoxy resin gel-casting process. Ceramics International, 2017, 43(7): 5478-5483.

[20] Guo H., Ye F., Li W., et al. Preparation and characterization of foamed microporous mullite ceramics based on kyanite. Ceramics International, 2015, 41(10): 14645-14651.

[21] Deng X.G., Wang J.K., Zhang H.J., et al. Effects of firing temperature on the microstructures and properties of porous mullite ceramics prepared by foam-gelcasting. Advances in Applied Ceramics,

2016, 115(4): 204-209.

[22]   Fukushima M., Yoshizawa Y. Fabrication and morphology control of highly porous mullite thermal insulators prepared by gelation freezing route. Journal of the European Ceramic Society, 2016, 36(12): 2947-2953.

[23]   Qian H., Cheng X., Zhang H., et al. Preparation of porous mullite ceramics using fly ash cenosphere as a pore-forming agent by gelcasting process. International Journal of Applied Ceramic Technology, 2014, 11(5): 858-863.

[24]   Hou Z., Du H., Liu J., et al. Fabrication and properties of mullite fiber matrix porous ceramics by a TBA-based gel-casting process. Journal of the European Ceramic Society, 2013, 33(4): 717-725.

[25]   Yang F., Li C., Lin Y., et al. Effects of sintering temperature on properties of porous mullite/corundum ceramics. Materials Letters, 2012, 73: 36-39.

[26]   Liu S., Liu J., Du H., et al. Hierarchical mullite structures and their heat-insulation and compression-resilience properties. Ceramics International, 2014, 40(4): 5611-5617.

[27]   Yang F.K., Li C.W., Lin Y.M., et al. Fabrication of porous mullite ceramics with high porosity using foam-gelcasting//Key Engineering Materials. Trans Tech Publications, 2012, 512: 580-585.

[28]   Wang Z., Feng P., Wang X., et al. Fabrication and properties of

freeze-cast mullite foams derived from coal-series kaolin. Ceramics International, 2016, 42(10): 12414-12421.

[29]   Dong X., Sui G., Yun Z., et al. Effect of temperature on the mechanical behavior of mullite fibrous ceramics with a 3D skeleton structure prepared by molding method. Materials & Design, 2016, 90: 942-948.

[30]   Zhang R., Hou X., Ye C., et al. Fabrication and properties of fibrous porous mullite–zirconia fiber networks with a quasi-layered structure. Journal of the European Ceramic Society, 2016, 36(14): 3539-3544.

[31]   Gong L., Wang Y., Cheng X., et al. Porous mullite ceramics with low thermal conductivity prepared by foaming and starch consolidation. Journal of Porous Materials, 2014, 21(1): 15-21.

[32]   Zhang R., Hou X., Ye C., et al. Enhanced mechanical and thermal properties of anisotropic fibrous porous mullite–zirconia composites produced using sol-gel impregnation. Journal of Alloys and Compounds, 2017, 699: 511-516.

[33]   Zhang R., Ye C., Hou X., et al. Microstructure and properties of lightweight fibrous porous mullite ceramics prepared by vacuum squeeze moulding technique. Ceramics International, 2016, 42(13): 14843-14848.

[34]   Wang Z., Feng P., Geng P., et al. Porous mullite thermal insulators from coal gangue fabricated by a starch-based foam gel-casting method.

Journal of the Australian Ceramic Society, 2017, 53(2): 287-291.

[35]   Li C., Bian C., Han Y., et al. Mullite whisker reinforced porous anorthite ceramics with low thermal conductivity and high strength. Journal of the European Ceramic Society, 2016, 36(3): 761-765.

[36]   Bourret J., Michot A., Tessier-Doyen N., et al. Thermal conductivity of very porous kaolin-based ceramics. Journal of the American Ceramic Society, 2014, 97(3): 938-944.

[37]   Sundaram S., Colombo P., Katoh Y. Selected emerging opportunities for ceramics in energy, environment, and transportation. International Journal of Applied Ceramic Technology, 2013, 10(5): 731-739.

6 General conclusions and outlooks

6.1 General conclusions

The accomplished study provides a strategy to prepare mullite foams with high geometric complexity, controlled pore structures and low thermal conductivity by Selective Laser Sintering (SLS) using fly ash hollow sphere (FAHS) as raw materials. The subject has a considerable practical interest because powder-based SLS is an emerging technology for manufacturing complex-shaped porous ceramic objects with great added value, and it also brings new sights into the advanced utilization of FAHS solid waste to help solve pollution problem.

1. Systematic study is accomplished for complex-shaped ceramic foams prepared by indirect SLS using FAHS as raw materials. The mechanical mixing and dissolution-precipitation methods are applied to prepare FHASs/PA12 composites with good flowability suitable for SLS experiment. The core-shell structured composites are prepared by dissolution-precipitation method, in which PA12 binder is coated on the surface of FAHSs. With increasing PA12 content, the average particle size of core-shell structured composites first increases and then decreases due to the self-crystallization and nucleation of PA12. Appropriate PA12 addition is ≤ 30 wt%.

2. Comprehensive experimental research on laser parameters are carried out for the fabrication of crack-free ceramic foams using C250 SLS

machine (Wuhan Huake 3D Technology Co. Ltd., China) equipped with CO_2 laser beam. SLS processing parameter combination, i.e., laser power of 6.6 W, scanning speed of 1800 mm/s and hatch spacing of 130 μm are selected based on the orthogonal experiment. Then crack-free honeycomb ceramic green body with intersecting holes is prepared using the above parameters.

3. TGA analysis is carried out to determine the heating cycle that contains two steps based on the data. The first is heating to 650°C for 2 h with a heating rate of 1°C/min to remove the binder phase. Then the temperature is increased to the expected sintering point (1250°C-1400°C) with a heating rate of 5°C/min and holding for 3 h to obtain the final ceramic foams. Calcined Al_2O_3 powders are used as a powder bed to provide good support and prevent the deformation or cracks throughout the heating cycle.

4. A bimodal pore structure could be observed in mullite foams prepared using FAHS as raw materials, i.e., closed pores from enclosed hollow spaces and open pores among FAHSs. The open pores of ceramic foams prepared by SLS technology mainly consists the interspaces between stacking FAHSs and the gaps between FAHSs directly related to SLS, which results in a higher total porosity in range of 79.9-89.1% of ceramic foams compared with that of foams prepared by conventional forming method.

5. The influence of sintering temperature on pore structure features (porosity, closed/open pore size distribution) and mechanical properties of ceramic foams is analysed. With increasing sintering temperature, the average closed and open pore size of mullite foams both decreases gradually, while the compressive strength increases dramatically. The strength enhancement of FAHS ceramic foams at low sintering temperature (<1350°C) is attributed to the reinforced sintering necks between spheres, while the densification and thickening of shell walls plays a key role when sintered above 1350°C. Furthermore, the fracture mechanism of FAHS foams changes from fracturing along spheres to across spheres with increasing sintering temperature.

6. The influence of PA12 content on phase composition, pore structure features and mechanical properties of ceramic foams is shown. The amount of mullite phase decreases with an increasing PA12 content, which is attributed to the reduced particle rearrangement and mass transfer rate. The open pores between FAHSs increase in the beginning due to the increased PA12 volume, and when the stacking volume of FAHS reaches the maximum, the excess PA12 will fill the limited gap between the FAHS causing the constant open pore channel size. Thermal conductivity of 0.08 W/(m·K) can be obtained when added 30 wt% PA12 content, which is lower than that of foams prepared by conventional forming methods due to the special pore structures in

FAHS ceramic foams prepared by SLS.

6.2 Further studies and developments

1. Further improvement of mechanical performance and the preparation of structural gradient material. Some sintering additives could be added into the composite powders to prepared ceramic foams with enhanced mechanical performance by SLS. Further efforts would be concentrated on fabrication of objects with variable and graded structures using FAHSs as raw materials.

2. Further improvement of high temperature resistance. At present, the mullite foams prepared by SLS show a rather low thermal conductivity, while its high temperature resistance requires further improvement considering the wide industrial application in exhaust filtration. Some additives with excellent high temperature resistance could be added to make possible processing ceramic foams with highly improved high temperature resistance.

More
Books!

OMNIScriptum

Printed by Books on Demand GmbH, Norderstedt / Germany